CW01278785

THE RESTORATION OF JEAN MacALPINE'S INN

(A Retirement Project)

By
Maggie and Sam Seabrook

Our efforts attracted an invitation to the Queen's Garden Party at Holyrood Palace and a Prince of Wales Award entry in 1994. We spoke to Prince Philip, but did not win the award.

THE RESTORATION OF JEAN MacALPINE'S INN

ISBN 978-1-907978-32-6

Published under the ETA BOOKS imprint by

ETA PUBLISHING HOUSE LLP Suite 2993, 6 Slington House, Rankine Road, BASINGSTOKE, United Kingdom, RG24 8PH.

Registered company number OC373475

Contents

Foreword

This book portrays the story of how Sam and Maggie restored a derelict Inn that they just happened upon whilst pursuing another hobby of theirs. It is amazing to me that out of the blue their hobby of 'collecting coloured model cottages' could result in such a valuable undertaking.

I can only imagine their surprise and joy when they realised that the ruins they had uncovered were that of 'Jean MacAlpine's Inn'. The very Inn or Cottar's Cottage so hugely featured in Sir Walter Scott's book 'Rob Roy' written so long ago in 1820.

At the time Walter Scott did the research for his book the Inn, situated in Milton, Aberfoyle, Scotland was already over 200 years old. It is almost as though this was meant to happen! No way was such an important building to be forgotten or left to rack and ruin.

The success of the 'Rob Roy' book led to it being made into a silent movie not once but three times between 1906 and 1922. Pictures of Jean MacAlpine's Inn not only appeared in the book but were also used as shots in the 1922 version of the movie. Understandably the Inn became very famous and this fame led to it having a major impact and contribution to the new trend of 'Tourism'. Tourism would bring the Railway, which in turn brought opportunity and new ways to make money to the area.

As I have indicated in what I have already written, Jeannie's Inn as it was to become affectionally called, was just too important to be forgotten.

As I read through the pages and shared in the journey with Sam and Maggie I could just feel how captivated they were by the story of the Cottar's Cottage and how it gripped their imagination, just as it had Sir Walter Scott's before them. At the same time they would surely have been astounded that yet another location, so rich with history and significance to the lives of so many, had been allowed to all but crumble away.

So they made a decision! Retirement was looming for both of them and restoring this cottage to its former glory would be an ideal retirement project. They retired, they bought the house and the garden containing the ruins. As they worked on the project they documented their progress with some 170 pictures, many of which are included in the pages to follow. These depict both the successful moments in sequence along with the different tasks undertaken. The pictures also show the difficulties encountered and how they were overcome. As I read on I was drawn into the project and inspired to do something similar myself.

As I considered not only the many other forgotten places of historical importance but also the old skills, dead or dying, like the buildings, I saw that this was not just a book; it is a legacy!

It is very evident that the restoration of Jean MacAlpine's Inn was a wonderful and enjoyable time for Sam and Maggie and the others involved along the way. In deciding to make this book available to all by publishing it through ETA Books, I believe that Sam and Maggie are leaving a wonderful legacy to those who will come after. Their depiction of the skills needed, now so sadly forgotten by most, are in these pages in wonderful colour, allowing for their rebirth too. I know you will enjoy this book, which in its own way is the full stop to this wonderful project completed over twenty years ago, as much as I have.

Cauline Thomas-Brown, DD

Author of Resting In Him

2nd May 2013

Chapter 1: History

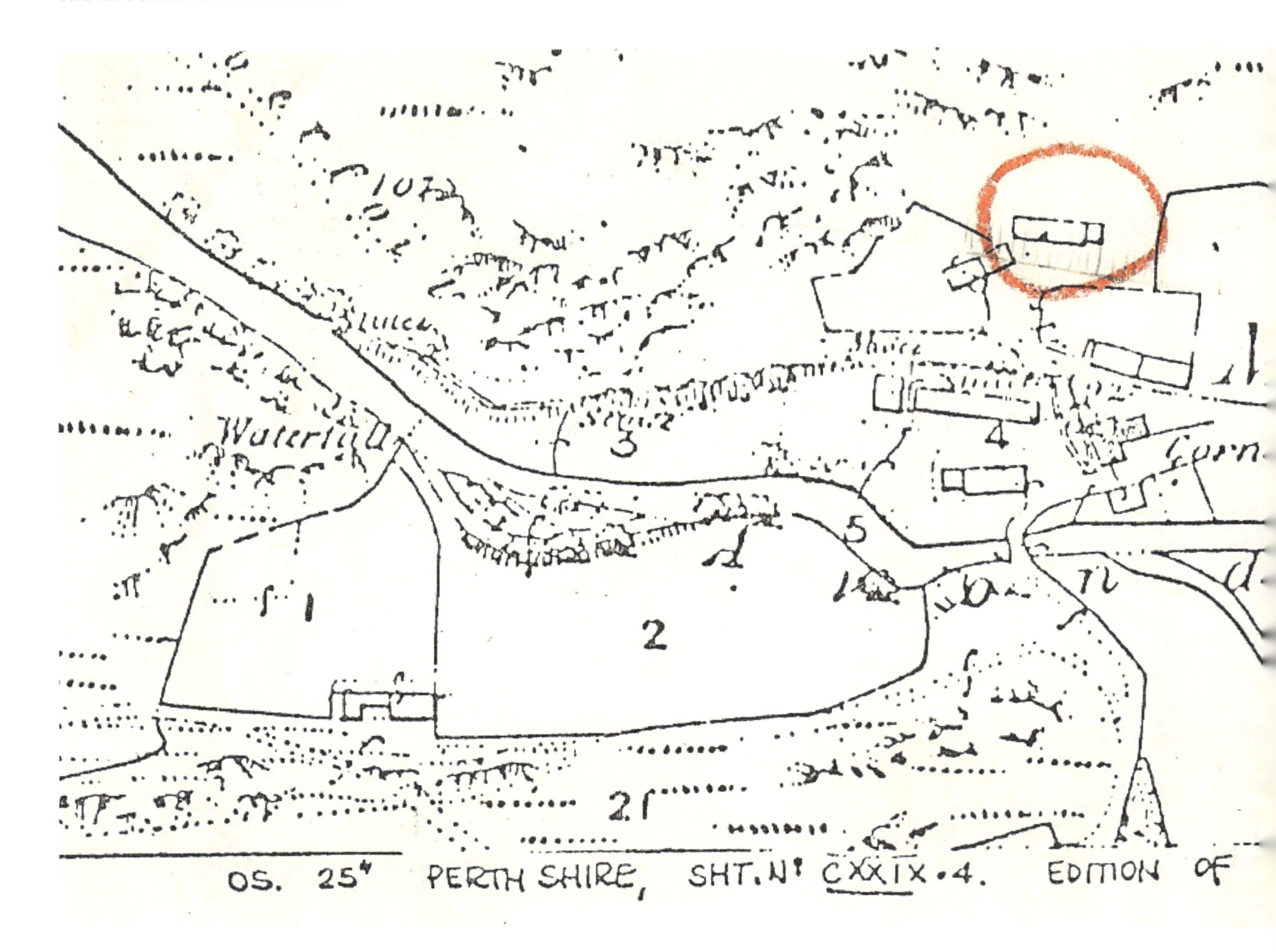

In 1986 when researching for our book on Collecting Miniature Coloured Cottages, ceramic souvenirs of the past, we came across the ruins of the original Jean MacAlpine's Inn, at The Milton in Aberfoyle, Stirling, Scotland.

The ruins are depicted on the pages before.

Jean MacAlpine's Inn, Arcadian Model circa 1925
4 inches across the front.

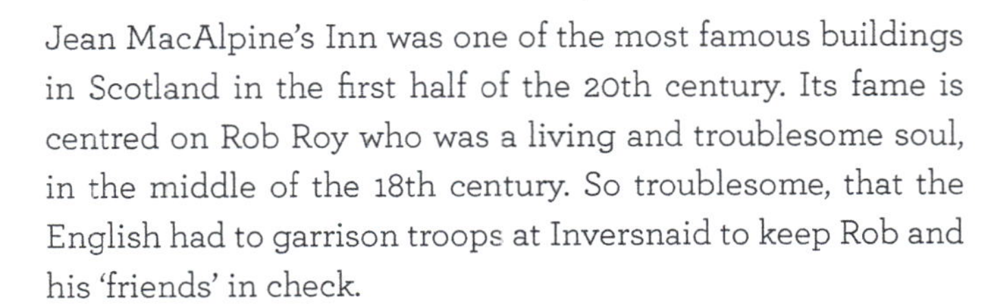

Jean MacAlpine's Inn was one of the most famous buildings in Scotland in the first half of the 20th century. Its fame is centred on Rob Roy who was a living and troublesome soul, in the middle of the 18th century. So troublesome, that the English had to garrison troops at Inversnaid to keep Rob and his 'friends' in check.

The community at The Milton and Jean MacAlpine's Inn were at the neck of an egg timer — everybody travelling North to Inversnaid or South to Stirling/Glasgow would have to pass (unless they went by boat over Loch Lomond).

Then in 1820 Walter Scott wrote the novel 'Rob Roy' where the scene of a fray between Bailie Nicol Jarvie and the Highlanders took place in Jean MacAlpine's Inn.

In 1922 the film Rob Roy featured Jean MacAlpine's Inn in many ways, and it was from this very early film that the Inn grew in popularity.

The potteries made coloured models of the Inn to be sold as souvenirs around 1925 — coloured china models, typically 2 — 4 inches wide were made of buildings associated with other Scottish writers — Robbie Burns, Walter Scott's (Fair Maid's, Perth), and J.M. Barrie (Peter Pan).

Potters also made coloured models of homes and/or buildings made popular by other British writers — Bunyan, Cain, Dickens, Hardy, Milton, Shakespeare, Walton, and Wordsworth.

Rob Roy was buried in Balquhidder, which is and was accessible with difficulty over the hills to Loch Katrine, Glengyle and Inversnaid, where he had properties.

Don't forget a decent road did not exist beyond The Milton (to Inversnaid) before 1855, or that many cottagers would produce their own whisky — and sell it.

The Pass of Aberfoyle to the left of Jean MacAlpine's Inn (circled red) was possible using many routes, depending on the state of the road and the amount of water. One of the ways was to pass Jean MacAlpine's Inn. (Some very nice houses have since been built on this land adjacent to the new road, in the area of the Pass of Aberfoyle).

Chapter 2: Site Clearance

We are not dry stone wallers/dykers but we know a man (Brian) who is.

We employed Brian 1 week in 4. Sam and I also went on courses for dry stone dyking and thatching ourselves at Glen Etive and Ruskie. It was a great learning curve. We attended farm sales to pick up the old tools needed, as we wanted it to be as authentic as possible. This was a fascinating time .

Our aim was to rebuild the Inn as it would have been in 1922, but in good condition and with no leaks.

Brian "what have I let myself in for!"

Jenny, our neighbour, a real enthusiast and a great supporter all the way through. Many Thanks.

The Inn was a ruin in the garden of Milton Farm Cottage, our house. We moved-in in 1988 when we retired from Hertfordshire and Essex County Councils. We had under estimated the size of the job — especially as we had no capital.

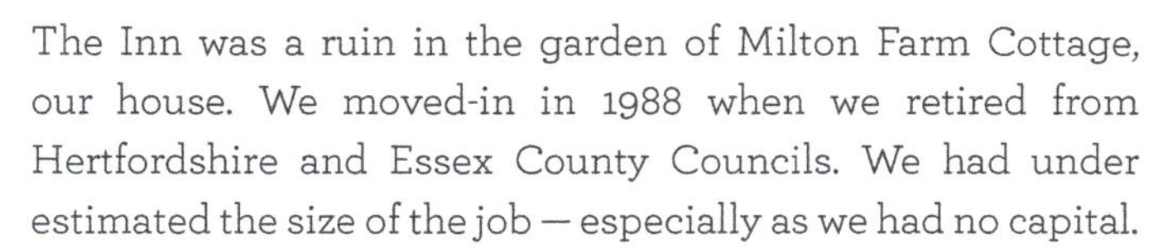

We started a Taxi Service '737' and Bed and Breakfast to bring in some income (both proved to be successful).

The chickens had to be rehoused, the new chicken house is behind the chap in the red bobble hat. He is finishing building the nesting box.

Our first aim was to rebuild sections one and two, that part of Jeannie's Inn on which the china model was based.

We rehoused the chickens and were then in the egg business particularly with regard to our B & B customers and a small amount of sales locally.

We cleared the first two units out, saving the best stone and revealing a very nice floor which we saved. We discovered an old fire place.

Latterly the old building had been used for dipping sheep and a tank (Fank) had been placed in the doorway for this purpose. The opening at the back was made for the sheep, we decided to keep the doorway. It is not usual for old cottages in Scotland to have a back door.

We needed to dig out the old tank (fank), we obtained ropes and dragged it out. We had a B & B guest with us, and she helped with the dragging, it was hard work, but we were a very happy crew when we finished. The tank still had dip in it also frogs, newts and other small water creatures surprisingly alive. The smell was obnoxious.

Jean MacAlpine's Inn. Existing Foundations & Projected Shape

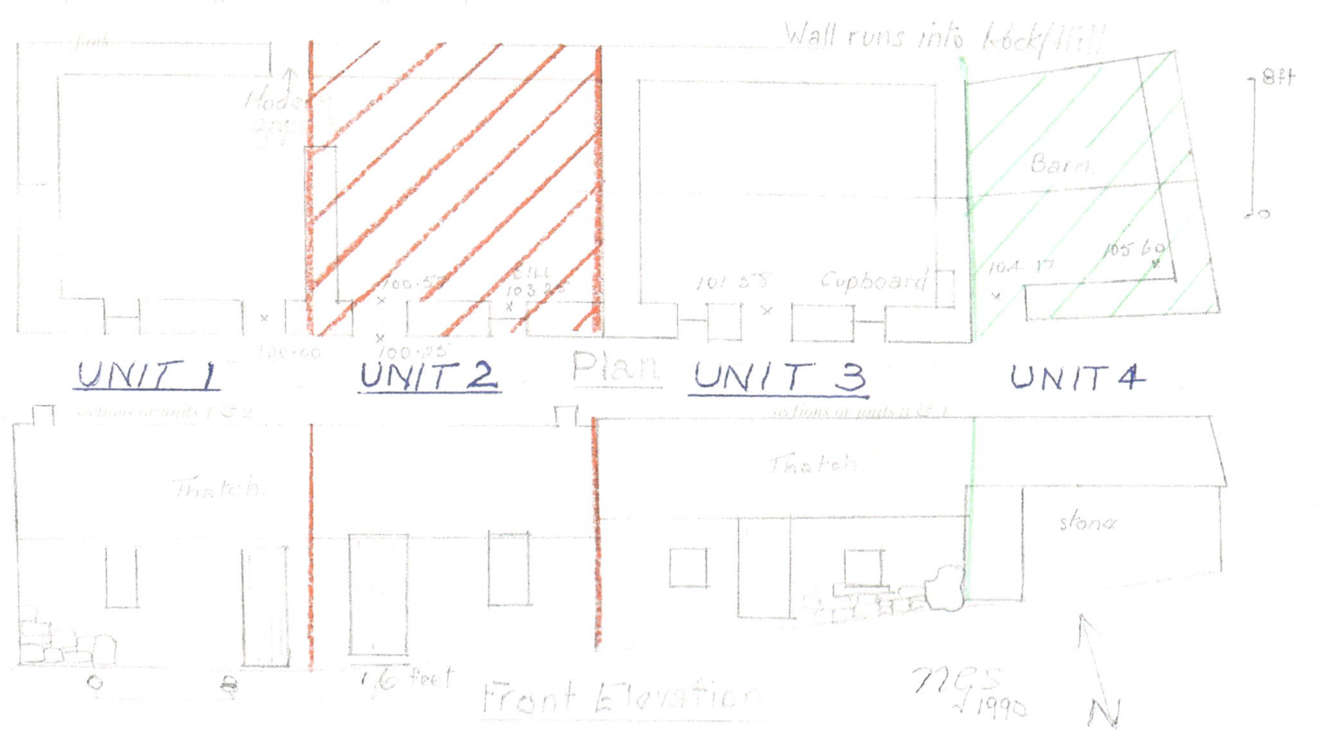

Jean MacAlpine's Inn around 1900.

Note rafters (fire) and deteriorating thatch to back of building. (Info. Jimmy Ferguson).

McKerracher with his Cuddie (donkey) at the front of the building.

Local families working the sheep at the back of Jean MacAlpine's Inn. (1909).

Jimmy Wright, who is often pictured on postcards etc., lived here in one room with a window (unit 2). He was still working as a stone breaker at the age of 74. He lived his life in Aberfoyle. Healthy Air. We were lucky!

An article appeared at this time in the 'Glasgow News' 15/07/1911, showing pictures of Jimmy Wright (then 84) with his dogs, 'Scotch Terriers', outside Jean MacAlpine's Inn, and he claims the house is over 300 years old (now over 400 years old).

Records indicate other people living here or about The Milton in thatched cottages around 1900. Children under 9 years were not always recorded at this time.

James (Jimmy) Wright 92 Died 1919

Flora McClean 85 Died 1894

Finlay Keir 72 Census 1901

Duncan McDonald 50 Census 1901

Elizabeth Menzies 88 Census 1901

A HIGHLAND CROFTER.

GIVING HIS DOGS AN AIRING.

["News" Photo.]

The old crofter pictured above is James Wright, of Milton, Aberfoyle, a well-known eder of Scotch terriers. The house beside which he was photographed is over 300 rs old.

CINEMATOGRAPH

IN THE CLACHA

ING "ROB ROY"

N OF ABERFOYLE.

Chapter 3: Old Filming History

Print on Photo (from the silent movie) reads, Killearn collects the rents for Montrose.

In 1911 United Films were making a mimic drama 'film' (i.e. without words) called 'Rob Roy' at the Clachan of Aberfoyle. There was a company of over 80 people (they were called cinematogragh records as they were in the open, also called mimic dramas).

In 1922, the Rob Roy silent film, featured the Inn. Jean MacAlpine was very proud of her adjacent doors— shown on the next page, one for the animals (beasts) and the other for customers, according to Scotts book (Rob Roy).

Bailie Nicol Jarvie defending the expected Rob Roy from would be captors, used a poker from the fire as a sword (as his sword was rusted into its scabbard). A kilt was set on fire, honour was restored, with everybody being involved in putting out the fire.

Many of the steps up to Jean MacAlpine's Inn are still in place and usable. (Picture from same old 1922 silent movie)

Four households lived here in the past with people sleeping in hammocks hanging along the walls. The house is in 4 units — the lower, in early times, was a barn in which 'beasts' were kept during the winter, the excrement could then drain away from the higher levels. Next the second section /unit would house 10 or more people some would sleep in the loft — this second unit was considered the main part of the house.

The third section is large and little is known about the human occupants, water flowed through the house and out the door when there was heavy rain. The fourth section (unit) may have been an outhouse though people did live in it.

Sir Walter Scott explored the Aberfoyle and The Milton area; Noting many anecdotes before writing his book 'Rob Roy'. Jean MacAlpine's Inn was in existence in the 18th century, maybe as a hostelry. It was used by farmers when they took their corn to be ground at the mill across the road.

In 1900, Jean MacAlpine's Inn was already 300 to 500 years old, and en-route from Aberfoyle to Inversnaid on Loch Lomond where Rob Roy had a cave and a house. At Loch Katrine Rob Roy had another house, Glengyle, where he was born, just over the hill from Balquhidder where he was buried.

In the 18th century the English built a garrison at Inversnaid, (the remains are still there) a young subaltern named Wolfe learned his trade trying to outwit Rob Roy. He became General Wolfe and later won battles at Quebec.

At Milton, the River Duchray and the Loch Ard outlet meet to create the source of the River Forth, which is tidal and was navigable as far as Stirling in the old days.

Chapter 4: Stone Building Work

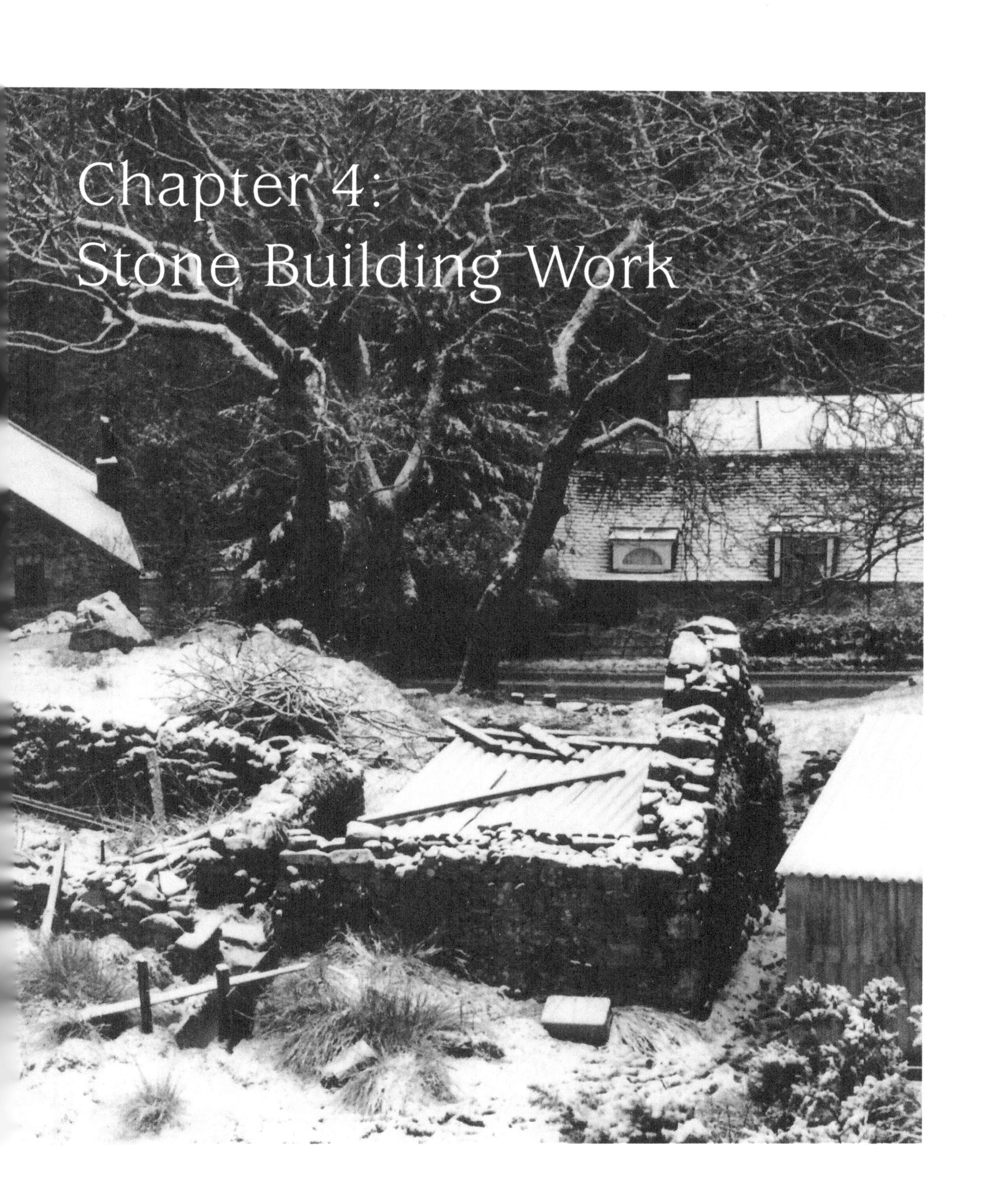

Extract from "Gazeteer of Scotland" Vol p.138.
The new road finished in the spring of 1855 is easily transversable by wheeled vehicles.
Previously access to Inversnaid was practicable only for pedestrians and ponies.

The plan was to start with the first two sections as shown on early postcards with Jimmy Wright at the front. This meant clearing the existing walls of fallen stone and rubbish so that, from the piles of loose stone, we could select the best for rebuilding.

This was quite interesting work as we were looking for stones that featured on the postcards of the Inn — we were singularly unlucky in this respect.

The Bailie Nicol Jarvie Hotel, so named in Victorian times was in existence before the films were made. It is in Aberfoyle at the junction of the old Drove Road (Dukes Pass) and Lochard Road about one mile from Jean MacAlpine's Inn at Milton.

The hotel no longer exists as a hotel, it has been converted into flats, and a nice group of small houses stand where the old stables stood. The 10ft wooden sign of the Bailie Nicol Jarvie is in safe keeping.

We needed to clear the site of loose stone and secure existing decent walls.

Cottage wall building differs from dry stone wall building (as around fields etc) in that the walls are more or less uniformly wide from top to bottom whilst field dry stone walls, are much wider at the bottom than the top.

The cottage dry stone walls are flat on both sides and usually the same width throughout.

The width of the wall must not get wider as it gets higher or it will become unstable.

Broadly each wall consists of two rows of stone, the front parallel to the back leaving a space between that will be filled with 'small stones' (hearting) as shown on page 24 — by Tweeny.

What is he thinking?

The 'Hearting' we used were of the same stone type as the main building stones – irregularly angular, about fist size, larger or smaller, and these were fitted into the many spaces between the parallel walls as they grew higher.

We decided that if the wall becomes a tiny bit narrower as the wall gets higher, it is stronger.

Tweeny helped.
Note – The line (string) guide for stone face.

Jenny supplying Brian with stone for wall.

Introducing our new Honda mechanical barrow. It's still going (2012) and was invaluable.

The heat from the sun encourages surface stones to expand and contract, the small hearting stones drop as the space is available making it unpredictable and unsafe.

''Hearting Stones' were used in vast quantities, gathered in old wire mesh potato baskets (now redundant with mechanised potato gathering).

String
The task of dry stone walling is a work of art and common sense.

Lime Mortar has been used in Britain since Roman times but was not always to hand when cottagers were building their cottage on their crofts miles from anywhere.

Avoid using successive upright joints; try to bridge an upright joint by placing a stone across it.

Before and After

We used Lime Mortar to stabilise the pillar between the two famous doors of Jean MacAlpine's Inn and the walls either side of the doors. Lime mortar takes different forms for different purposes — we bought ours already prepared from a firm in Leith, near Edinburgh.

Some stones span the width of the wall resting on both the inside and outside walls — these are called 'throughs' and are invaluable as stabilizers.

A small stonemason's hammer can be used to shape some stones (large and small).

Chapter 5: Woodwork – The

Beams were needed to span the width of the building and to construct the roof frame — a massive job.

We looked around and found one or two but eventually sought the advice of the Forestry Commission who pointed us to Loch Awe (North side). A 160 mile round trip, where armed with red paint, a list and a measuring tape we marked the tree trunks we needed.

Brian and his Dad

The tree trunks were already stacked in huge piles and the Forestry Commission said we could have what we wanted free; but we had to pay for the transport which we organized. Eventually the tree trunks arrived in two deliveries. The whole operation went smoothly.

The wood preparation for the building had now begun — we started adzing and splitting the tree trunks where necessary. All the trunks had to be adzed (hard work).

Splitting the heavy adzed beams along the grain using iron wedges, axe and sledge hammers.

North Wall (3rd Unit) Looking from South.

Here, at the Eastern end the walls are not mortared and couples or crucks are required to support the roof. These sit on base stones then the dry stone wall is built around the upright.

Below, at the Western end some walls are mortared and can support an 'A' Frame.

Brian can be seen constructing the hanging 'lum' or chimney breast, which will be made from Willow and Hazel branches covered with clay and lime wash. (Page 31)

View from back of Inn (from north).

Dowels were made on these benches using a draw knife.

Suitable oak off cuts from the beams were chopped into approx. one and a half inches square lengths about 12 to 18 inches long, then rounded off.

Cruck beam joint with dowel and reel.

Is this Kerrie Foot with Jenny Legge and Marjorie Mclean making dowels?

Putting another dowel in the main cruck beam joint.

The two cruck beams are joined at the top.

The picture shows the ridge pole in place supported at one end by a ladder.

The cruck beams are raised, as one, already joined and with dowels in place.

Hope Rowan is not going to cut the Ridge pole (unit 1)

Now we can concentrate on the first and second units again; obviously there is an overlap between 2nd and 3rd units with the common wall supporting both roofs and that is why the couples needed to be erected first.

All we needed to do now that the roof structure was in place was (1) source, collect and erect cabers (+tie), (2) source and collect turf, (3) source and collect thatch, turf ridge and thatch.

But first we had to get the lum (chimney) in place.

There is no a window in the original Jean MacAlpine's Inn on the north wall.

This is not a picture of Jean MacAlpine's interior though similar.

Crofter's Cottage, Interior

Chapter 6: Woodwork – The Cabers

This is early 1992 and we were hoping to thatch the roof of sections 1 and 2 this year.

We sought permission from the Forestry Commission to thin parts of the forest along Bell House Road beyond the Covenanters Hotel and borrowed a tractor and trailer from John Mclellan (local farmer).

Nearly 200 x 11ft plus cabers were required for the first two sections and all had to be stripped of bark and treated before being positioned and cut to length.

The cabers were sent away to be treated, as this would preserve their life (this was the only variation from the old traditional ways).

Nearly 200 x 11ft plus cabers were required for the first two sections and all had to be stripped of bark and treated before being positioned and cut to length.

The cabers were sent away to be treated as this would preserve their life (this was the only variation from the traditional ways).

Also below — see nice stone lintels above door and window.

We raised the heaviest lintels by first placing a heap of stones across the doorway, then by 4 to 6 inches moves, gradually raised the lintels using the existing walls for support and by tucking stones under each side until we reached the required height, then sliding the lintel into place.

Always standing well clear we decided there must be a better way of doing it. It had taken 4 of us ages (well, all afternoon).

Cabers in position on the first 2 sections (north side) but also note top left, the woven lum 'breast' which will later be covered with clay and lime

Note — the cabers are tied with specially treated twine, see ridge internally and bottom right of straight cabers.

Whatever we were doing (splitting a beam?) we are doing it in the shelter of the Inn. Note

1. The three new bib and brace doors are in place – professionally made in Stirling but painted by us. The doors were placed at this time to keep out the wildlife including bats.
2. The top of the lum has been covered in clay and lime wash finished.
3. The cabers have been tied to the ridge (no nails).
4. The planks are an aid for construction and are loose in a place where people would have slept.

We produced the yellow circular, shown in the next chapter, which was circulated through Scottish Conservation Projects.

Chapter 7: Bracken Pulling

THATCH-GATHERING WEEKEND
24th-28th SEPTEMBER 1992

VOLUNTEERS WANTED FOR TRADITIONAL BRACKEN-PULLING/SOCIAL WEEKEND."RETURN TO THE NINETEENTH CENTURY"
FOR FINAL PHASE OF RESTORATION OF HISTORIC
"JEAN MacALPINE'S INN" COTTAGE.

ALL MEALS PROVIDED.
TELEPHONE
MAGGIE OR SAM ON
087 72 737
FOR DETAILS

Inversnaid.
Kinlochard,
Aberfoyle
MILTON
Stirling 20ml
A873
A81
Jean MacAlpine's Inn
Milton
By Aberfoyle
Glasgow 30ml

PLEASE BRING - WATERPROOFS, GLOVES, SLEEPING BAGS
ACCOMMODATION & CAMPING. CAN BE ARRANGED-PLEASE TELEPHONE IF REQUIRED.

SHELL BETTER BRITAIN CAMPAIGN -AWARD WINNER.

The picture below shows sheaves of bracken — some of the first pulled — with the cabers in place on the south side and the scaffolding — all before turfing.

Now for the thatching — we needed lots of volunteers to help to pull bracken for the main two sections; fortunately we received a grant from Shell Better Britain for £800+ to help feed everybody and for protective gear (face masks and gloves). We had eighty volunteers approx. who slept on site, in lofts, tents, on the floor of the house and in cars. We had lovely weather throughout. We were lucky!

We employed Duncan Matheson and his son Doodie from Kintail on the recommendation of Jim Souness, Duncan is pictured here with Doodie. Duncan and Doodie stayed with us for three weeks laying the turf then thatching.

This is the back of the Inn (north facing side). The turf arrived; it had first been laid on an old building which was being demolished, luckily someone thought of us and preserved it. The turf was about 200 years old and was rather like old pieces of carpet made from sisal.

Now the roof was ready for thatching.

It was the first time bracken had been used for thatching for more than 50 years, quite an undertaking and challenge for Duncan and his son — exciting!

Pulling the bracken was hard work - it is called 'pulling' as the black root (the water proof part) of the bracken had to be pulled to the 4-6 inches we needed.

Edinburgh volunteers including Willy.

More Scottish Volunteers

Back from Hatfield

The more the merrier (Fraser and family)

Edith defronding with the scissors.

Two pairs of hands better than one?

Catriona.

Ruaridh (Rory)

Brian with Jim Souness.

Chapter 8: Bracken Thatching

415 bundles or sheaves of bracken were transported from Aberfoyle, broken up into groups of Dollies, then laid on the turf.

The Dollies were made on the Dolly Frame both shown below. As the size depended on requirements some where made manually. Only the weatherproof black root was exposed to enable the Dollies to last for the 100 years lifespan claimed. All the bracken was used on both sides of the first two sections of the Inn. All this was possible, thanks to the Pullers' who came from all over the world including Australia and the USA. Everyone worked as a team, it was great!

We will leave it to the experts Duncan and Doodie Matheson.

We hired the scaffolding from a regular scaffolding firm, who erected it as indicated, so that we could work on the thatch. They charged us a weekly rate, very reasonable.

Note the turfs for the ridge on the ground in picture below.

Duncan and Dolly

Long hours when the weather allows.

Black and white photographs courtesy of John Kerr.

Thatch progressing on north side. Note wire on right.

As the bracken thatch progressed to the ridge, secured at intervals by wire, the waterproof problem at the ridge was solved by placing a long turf over the top i.e. straddling the top — green side up — the weight of the turf holding it in place.

Brian cutting turf for Ridge using turfing iron, whilst thatching is in progress.

Rolling Turf in 5ft Lengths, and Transporting turf.

The roofs of the first sections were then finished except for the turf ridge. A plastic sheet was needed to protect the lovely bracken thatch finish from straw and turf debris. In addition to this speed was of the essence to ensure that the thatch did not deteriorate quickly so we had to work both sides of the roof at once.

A small amount of straw is packed along the top of the ridge pole to support the turf. Note the straw on the scaffolding.

The turfs were heavy and two ladders and two people (Doodie and Rowan) were needed to carry the turf to the ridge. (I, Sam, supervised!). With this done we were able to move on to the final two sections.

Chickens making eggs.

A few wooden pegs required in turf below Fairy Chimney.
The Fairy Chimneys were constructed with bracken and sisal (rope)).

We can move on to the final two sections, but not without mentioning the fairy chimneys, which are bracken and sisal (rope).

Scots Magazine Newsagents advert Aug/Sept 1994 actual size 14 x 10.5 inches. Nice article by John Harrison.

ABERFOYLE Dream Cottage

The restoration of a house made famous by Sir Walter Scott — Jean MacAlpine's Inn.

Chapter 9: The Project So Far

We wrote an intermediate report in March 1994 for the Goss Club, as we are members, explaining that we are restoring Jean MacAlpine's Inn using the tools and techniques used in those times where possible. We had now thatched the first and second sections and could comment on the progress and work so far.

We had learnt (apart from acquiring an ability to disregard inclement weather) that the amount of work/manpower hours and general input required was unbelievable!

How people could bring up a family, work, eat and keep warm (doubt if they did!) without state benefits, modern tools or a mechanised barrow (fabulous)? How did they also manage ridge poles, special lintels or enough material to thatch their homes? It must have been soul destroying!

Financially we managed to pay 80-90% of the cost ourselves with the remainder coming from a couple of other organisations.

The physical help from the volunteers and professionals was hard work but proved enjoyable for all. For nearly 6 weeks we had sun and for 3 weeks we had 80 volunteers to pull bracken, good fun and lucky with the weather! We also employed experts to thatch the roof.

We have set up a group to advise "The Jean MacAlpine's Inn Restoration Group" comprising of ourselves, Sam Scott, Royal Commission on the Ancient Historical Monuments of Scotland, Jim Sounness of Friends of Thatched Houses, Rowan Seabrook, Brian Wilson and other experts as required e.g Martin Hadlington for Lime Mortar and Frank Bracewell for good advice together with Veronica Howard representative of the Scottish Conservation Project Trust.

Chapter 10: The Third and Fourth Sections

All the stone work was complete on the first two sections.

The gable end, oak beams, purlins, etc. were in position on the third and fourth sections, cabers were cut, debarked and treated. The heather turf was being gathered, but first we had to collect and position the cabers. Then wait for the spring when we would gather the broom when ready, our next thatching material.

In the meantime Brian and Rowan were renovating Moirlanich Long House near Killin on behalf of the Scottish National Trust.

Chapter 11 : The Cabers and Turf

These are the third and fourth units. Note the Broom had started arriving for the third unit, it is leaning on the wall of the fourth unit ready to be used.

We now proceeded to cut the turf by Loch Katrine with the permission of the Water Board.

We cut the heather turf to shape where possible, because it has to be transported by car the 10 miles to The Milton and there is little point in moving stuff that is to be thrown away.

Other turf came from our sheep field, thanks to John McClellan, which could be easily transported on the 'Honda'.

We think this is Russell by Loch Katrine and know it is Rowan on the next page at the Milton.

Seven people all working at once (is this a British record?). Shaping turf for the roof and ridge — transporting broom and laying turf.

The small picture on the right shows the height of the broom before cutting.

Chapter 12: The Brooms and Pegs

Broom sheaves!

Willy Dewar — with broom (he was instrumental in getting the Edinburgh Post Office workers to volunteer). Turf in place and broom ready to be used as thatch — all on the third section.

A group of chaps from the Royal Mail, Edinburgh and others helped to cut the broom for the third section — The broom from Thornhill and Doune (Doune Broom)

See the yellow string behind the van.

The broom had to be fixed with pegs — Rowan can be seen on the next page making pegs from hazel sticks. They were cut to length, sharpened, then heated and twisted before being pushed into place over the broom.

No pegs yet!

Roncan using the original grate.

The hazel sticks, were gleaned from the forest perimeter and cut to lengths, those used as pegs 2ft—4ft long were sharpened at both ends and heated over the fire so that they were easily twisted.

Later they were pushed between the broom and cabers or positioned to hold longer lengths of hazel in place and fixed on the other side of the thatch or cabers.

The hazel was used in all sorts of forms and in all sorts of places to help fix the broom e.g. where there was a difference in level as between bracken and broom.

Either side of the thatch we had Brian and Rowan stitching and fixing the broom with needle and tar twine or wire.

Brian putting broom in place before fixing. Sheaves first positioned and pegged, then overlaid with smaller quantities and fixed as shown.

The third section needed only the turf over the ridge — and then some internal embellishment. Now we were able to move onto the final section whenever we were ready.

Note — Some of the hazel pegs were fixed with tar twine e.g. As displayed by Rowan in the picture below.

The fourth unit, the oak framework, the cabers and the shaped turfs all in place (nice aren't they!) after which the straw thatch and then people were asking us if they could use it for filming.

Final thatch, straw, shown on first of 'Alive and Kicking' pictures.

Chapter 13: The National Park

A lady with vision and ideas, Pat Macinnes, attempted to use The Inn by turning it into "a living History Centre". Her idea was using and teaching/training interested folk about dyes and natural materials from around Milton, all in a resurgence of old skills but sadly her ideas did not materialise, 'though she did manage to get the building listed, a preservation order grade Category B was issued'. We were sorry the idea came to nothing; but the B category listing persists.

Thank you and well done!

Jean MacAlpine's Inn is within the boundaries of The Loch Lomond & The Trossachs National Park which was the first National Park in Scotland and only established in 2002, after we had finished our work. Jean MacAlpine's Inn may, therefore, be preserved for all time!!!

Chapter 14: Modern Use

This picture shows me (Maggie) in 'Alive and Kicking' as Mrs Burns (Robbie Burns mum) trying to drive the sheep, (Daffy and Timmy, the Jacob) around the cottage – without success as they wanted to follow me, as normal.

'Alive and Kicking' a story of Robbie Burns was filmed in a large part here; they used the building as his cottage, Maggie as his mum, me as a customer at the Inn, Tom Watson as orator and the upper part of unit three as Burns' office.

Mrs. Burns (Maggie)

The horse was used in the story 'Tam O' Shanter' and looked quite eerie in the dark when racing along on the forest roads

The above pictures are all part of the 'Alive and Kicking' 'Series. which was shown in schools.

Rob Roys favourite pub being entertained by the Wallace clan, dressed in various clan colours, waving swords about and overall everyone enjoying the spectacle. The show was for entertaining pictures to help in selling the old/new Jean MacAlpine's Inn all taken for The Sun by The Sun's photographers.

Just having some modern day fun!

A picture of the Wallace clan inside Jean MacAlpine's Inn, the date on the newspaper is 20th May 1995.

The newly rebuilt cottage, as here, also had a fire place, lum and bible shelves. The new owners removed all three replacing the bible shelves with a door into the third unit. A door was necessary.

Interior, Highland Cottage, Aberfoyle.

Chapter 15: The Sale and Present Use

THE MIRROR, Tuesday, May 27, 1997

Rob Roy pub sale

ROB ROY MacGregor's favourite pub – Jeanie MacAlpine's Inn – has been put up for sale.

The 500-year-old boozer at the Clachan of Milton in Aberfoyle was a regular haunt of the legendary Highland raider.

Now the pub, which sits at the foot of Craig Mhor near the Trossachs, is for sale at offers over £50,000.

After renovations of Jeannie's Inn were finished we decided to move to Calne, Wiltshire. We sold the house we owned on the land but retained Jean MacAlpine's for a while. During this time we introduced storage heating with a view to getting planning permission to develop Jean MacAlpine's as a house, this took some time, but we were successful in the end. Armed with planning permission we decided to sell.

The new owners completed the development by joining the rooms internally with the introduction of doorways. They also converted the second unit into a kitchen, did away with the old lum and added a bathroom complete with window to the back of the house.

Externally, they rethatched the whole house with Tay Reed leaving the old thatching in place, removed the telegraph pole by unit 4 and provided four lovely glazed doors that they painted white.

Jeannie's Inn is now a delightful family home and a far cry from what it was. With today's occupants enjoying all the comforts of modern day living and non of the hardships of latter years endured by its original occupants.

'Take The High Road' which was a very popular series was also filmed here. The usual filming location was on the other side of Loch Lomond, but for one particular story line, where a suitable place/bothy where Sneddon could seduce a girl – Jeannie's Inn was ideal. During filming the chickens clucked continuously and rather loudly resulting in retakes. A nightmare for the film crew but rather humorous for us at the time.

Before

After

From this 1996...

Most of the photos were taken by Maggie but the 3 most recent pictures were by courtesy of Dr. & Mrs. C. Stokoe.

...to this 2011.

Authors Biographies

Maggie and Sam are both Octogenarians who have been married for over thirty years. Their twinned experience gained in local government and covering building design, building construction, extensive report writing and hobbies have formed the perfect basis for their writing career.

The couple live in Calne, Wiltshire and have five sons, nine grandchildren and three great grandchildren.

Woolly on watch!

Acknowledgments

Pamela Abrams

Ingrid & John Anderson

Sean Barrington

Sue Beech

Frank Bracewell

Willy Dewar

R & N Fowler

Catriona Harris

Janet Harris

John Kerr

Margaret Latham

Jenny Legge

Doodie Matheson

Duncan Matheson

Marjorie Mclean

John & Toto McLellan

Pat Macinnes

Chris and Rae Moorey

Sam Scott

Jim and Alan Scott

Rowan Seabrook

Mathew Seabrook

Jim Souness

Brian Wilson

and

To the hundreds of other volunteers and family members, ladies and gentlemen, boys & girls for their help and humour.

Thank you to everyone.

Visit: www.etapublishing.com

Write to: ETA PUBLISHING HOUSE LLP Suite 2993, 6 Slington House, Rankine Road, BASINGSTOKE, United Kingdom, RG24 8PH

Email: info@etapublishing.com

Telephone: 0843 289 2274

Fax: 0871 277 3138

Lightning Source UK Ltd.
Milton Keynes UK
UKOW06n1423310713

214675UK00003B/6/P

9 781907 978326